Electrical Systems in the Human Body

By Ashai J D Parsons

ISBN: 9798392981786

DEDICATION

To Georgie Bark, a cherished friend who never failed to electrify every gathering we attended, and whose vibrant energy continues to illuminate every path she takes.

CONTENTS

Electrical Systems in the Human Body

The human body is a complex and remarkable system that exhibits an array of intricate biological and chemical processes. Among these, the electrical systems within the human body have drawn significant attention due to their critical role in various functions such as nerve impulse transmission, muscle contraction, and sensory perception. This scientific investigation aims to explore the underlying mechanisms, properties, and interactions of the electrical systems in the human body, delving into their role in maintaining homeostasis and their potential applications in medical and biotechnological advancements.

BACKGROUND AND CONTEXT

The electrical nature of the human body has long fascinated scientists, philosophers, and the general public. From ancient times, when the first sparks of curiosity were ignited, to the ground-breaking discoveries of pioneers like Luigi Galvani and Alessandro Volta, our understanding of the body's electrical systems has expanded tremendously. These early investigations laid the foundation for modern research in electrophysiology, neurobiology, and numerous other fields. In this book, we delve deeper into the intricacies of the human body's electrical systems, exploring the physiological processes in which electrical signals play a crucial role.

As you journey through the chapters of this book, you can expect to gain insights into the following aspects of the human body's electrical systems:

> Historical perspectives: We begin by exploring the early discoveries and key milestones that shaped

our understanding of the body's electrical systems. From Galvani's experiments with frog legs to Volta's invention of the battery, we will examine how these early breakthroughs paved the way for future research.

Cellular mechanisms of electrical signals: We will dive into the fundamental mechanisms that generate electrical signals in cells, particularly neurons and muscle cells. This includes an in-depth exploration of the roles of ion channels, pumps, and other key components involved in maintaining electrical potentials across cell membranes.

Signal transmission and propagation: The book will discuss how electrical signals are transmitted along neurons, including the process of action potential generation and propagation. We will also explore the role of electrical signals in the transmission of information between neurons through synapses.

The nervous system and sensory perception: We will examine how the electrical systems in the human body contribute to sensory perception, muscle contraction, and overall function. This includes a detailed analysis of the roles of various types of neurons, the sensory receptors that detect environmental stimuli, and the neural circuits responsible for processing and integrating sensory information.

Biomedical applications and future directions: As we venture into the potential applications of electrical system research, we will explore how our growing understanding of these systems can be harnessed for medical advancements. This includes the development of innovative technologies such as neuroprosthetics, brain-computer interfaces, and other cutting-edge therapies for neurological and muscular disorders.

Interdisciplinary research: Throughout the book, we will emphasize the importance of interdisciplinary research in deepening our understanding of the human body's electrical systems. We will discuss how various fields, including biophysics, electrophysiology, molecular biology, biochemistry, neuroscience, and engineering, can contribute to a more comprehensive understanding of these systems and their potential applications.

By providing a comprehensive analysis of the electrical systems within the human body, this book aims to contribute to the current understanding of the fundamental principles governing these systems and their role in maintaining overall health and function. Furthermore, it seeks to inspire further exploration and pave the way for novel medical applications and technologies, ultimately leading to improved diagnostic and therapeutic options for patients suffering from various neurological and muscular disorders.

ELECTRICAL MECHANISMS IN NEURONS AND MUSCLE CELLS

The generation of electrical signals in cells, particularly neurons and muscle cells, is a crucial aspect of the human body's electrical systems. These signals, which result from the flow of ions across cell membranes, are responsible for transmitting information throughout the body and facilitating various physiological processes such as muscle contraction and sensory perception. In this section, we will explore the fundamental mechanisms that generate electrical signals in neurons and muscle cells, focusing on the concepts of resting membrane potential, action potentials, and the role of ion channels and pumps.

Resting Membrane Potential

The resting membrane potential (RMP) refers to the difference in electrical potential between the inside and the outside of a cell at rest. This potential difference, which is typically around -70 millivolts (mV) in neurons, is primarily due to the unequal distribution of ions across the cell membrane, with higher concentrations of potassium (K+) ions inside the cell and sodium (Na+) ions outside. The RMP is maintained by the selective permeability of the cell membrane to specific ions and the activity of ion pumps, such as the sodium-potassium (Na+/K+) ATPase pump, which actively transports three Na+ ions out of the cell and two K+ ions into the cell, maintaining the ion concentration gradient.

Action Potentials

An action potential is a rapid, transient change in membrane potential that propagates along the cell membrane, allowing for the transmission of electrical signals in neurons and muscle cells. The generation of an action potential involves a series of well-orchestrated events, including:

Depolarization: Stimuli, such as neurotransmitters or mechanical forces, can cause a localized increase in the permeability of the cell membrane to Na+ ions, resulting in an influx of Na+ ions and a decrease in the membrane potential (i.e., depolarization). When the membrane potential reaches a critical threshold (usually around -55 mV), voltage-gated Na+ channels open, causing a rapid influx of Na+ ions and a further depolarization of the membrane.

Repolarization: As the membrane potential approaches +30 mV, the voltage-gated Na+ channels begin to inactivate, while voltage-gated potassium (K+) channels open, allowing K+ ions to flow out of the cell. This efflux of K+ ions cause the membrane potential to become more negative, leading to repolarization.

Hyperpolarization and Return to Resting Membrane Potential: The efflux of K+ ions can momentarily cause the membrane potential to become more negative than the resting membrane potential, resulting in hyperpolarization. Subsequently, the ion concentrations are restored to their resting state by the Na+/K+ ATPase pump and the passive diffusion of ions through leak channels, re-establishing the resting membrane potential.

ION CHANNELS AND PUMPS

Ion channels and pumps play a crucial role in the generation of electrical signals in neurons and muscle cells. Ion channels are transmembrane proteins that selectively allow ions to pass through the cell membrane, either passively through leak channels or in response to specific stimuli, such as voltage changes or the binding of ligands (e.g., neurotransmitters). Voltage-gated ion channels, such as the Na+ and K+ channels mentioned earlier, are particularly important for the generation and propagation of action potentials.

Ion pumps, on the other hand, actively transport ions across the cell membrane, using energy from ATP hydrolysis to maintain the ion concentration gradients necessary for the resting membrane potential and the generation of electrical signals. The Na+/K+ ATPase pump is a prime example of an ion pump that plays a critical role in maintaining the ion gradients and resting membrane potential in both neurons and muscle cells.

Electrical Signal Generation in Neurons

In neurons, electrical signals are responsible for transmitting information throughout the nervous system. The generation of action potentials in neurons follows the same fundamental principles discussed earlier, with some unique features:

Graded potentials: Before an action potential is generated, neurons can experience graded potentials, which are small, localized changes in membrane potential that can either be depolarizing or hyperpolarizing. These graded potentials are caused by the opening of ligand-gated ion channels in response to neurotransmitters binding at the synapses. If the graded potential is depolarizing and reaches the threshold, it can trigger an action potential at the axon hillock, where the density of voltage-gated Na+ channels is high.

All-or-none principle: Neuronal action potentials follow the all-or-none principle, meaning that once the threshold is reached, a full action potential is generated, and its amplitude remains constant as it propagates along the axon. If the threshold is not reached, no action potential is produced.

Refractory periods: Neurons exhibit refractory periods, during which they are either incapable (absolute refractory period) or less likely (relative refractory period) to generate another action potential. Refractory periods ensure the unidirectional propagation of action potentials along the axon and prevent the continuous firing of action potentials.

Electrical Signal Generation in Muscle Cells

In muscle cells, electrical signals play a vital role in initiating muscle contraction. The process of electrical signal generation in muscle cells, specifically skeletal muscle cells, is similar to that in neurons, with some distinctions:

Excitation-contraction coupling: In muscle cells, the generation of an action potential is closely linked to the process of muscle contraction through a mechanism known as excitation-contraction coupling. When an action potential reaches the muscle cell via a motor neuron, it triggers the opening of voltage-gated L-type calcium (Ca2+) channels, allowing Ca2+ ions to enter the cell. This influx of Ca2+ ions initiate a series of events that ultimately leads to the release of calcium from the sarcoplasmic reticulum, an intracellular calcium store, and the subsequent contraction of the muscle fibre.

Muscle cell action potential propagation: Unlike neurons, where action potentials propagate along the axon, muscle cell action potentials propagate along the sarcolemma (cell membrane) and down the transverse (T) tubules, which are invaginations of the sarcolemma that penetrate deep into the muscle fibre. This allows for the rapid and simultaneous activation of multiple muscle fibres within a muscle, resulting in coordinated muscle contractions.

Thus, the fundamental mechanisms that generate electrical signals in cells, particularly neurons and muscle cells, involve the establishment and maintenance of the resting membrane potential, the generation and propagation of action potentials, and the crucial roles played by ion channels and pumps. Understanding these mechanisms is essential for gaining insights into the complex electrical systems of the human body and their roles in various physiological processes.

The Role of Ion Channels and Pumps in Maintaining Electrical Potential Across Cell Membranes

The electrical potential across cell membranes, which is critical for various physiological processes, is primarily maintained by the interplay between ion channels and pumps. These specialized proteins are responsible for regulating the movement of ions across the membrane, thereby establishing, and maintaining the resting membrane potential and facilitating the generation and propagation of electrical signals. In this section, we will discuss the specific roles of ion channels and pumps in maintaining the electrical potential across cell membranes

Ion Channels

Ion channels are transmembrane proteins that form selective pores, allowing specific ions to pass through the cell membrane either passively (down their concentration gradient) or in response to specific stimuli. Ion channels can be categorized based on their gating mechanisms, which determine when the channels open or close:

Leak channels: These channels are open at rest, allowing for the passive movement of ions across the cell membrane. Leak channels contribute to the resting membrane potential by maintaining a low level of ion permeability. For instance, potassium leak channels permit the passive diffusion of K^+ ions out of the cell, making the membrane potential more negative.

Voltage-gated channels: These channels open or close in response to changes in the membrane potential. Voltage-gated Na^+ and K^+ channels play a crucial role in the generation and propagation of action potentials in neurons and muscle cells. Voltage-gated Ca^{2+} channels are particularly important in muscle cells, where they are involved in excitation-contraction coupling.

Ligand-gated channels: These channels open or close upon binding of specific ligands, such as neurotransmitters or other signalling molecules. In neurons, ligand-gated channels are essential for generating graded potentials in response to neurotransmitter binding at synapses.

Ion Pumps

Ion pumps are membrane-bound enzymes that actively transport ions across the cell membrane against their concentration gradient. This process requires energy, usually in the form of ATP hydrolysis. Ion pumps help maintain the electrical potential across cell membranes by establishing and preserving the concentration gradients of specific ions, which are essential for the resting membrane potential and the generation of electrical signals. Some prominent ion pumps include:

Sodium-potassium (Na^+/K^+) ATPase pump: This pump plays a vital role in maintaining the resting membrane potential in both neurons and muscle cells by actively transporting three Na^+ ions out of the cell and two K^+ ions into the cell for each ATP molecule hydrolysed. This process helps establish the concentration gradients of Na^+ and K^+ ions, which are critical for the generation of electrical signals.

Calcium (Ca^{2+}) ATPase pump: This pump actively transports Ca^{2+} ions out of the cell or into intracellular compartments, such as the sarcoplasmic reticulum in muscle cells. By maintaining low intracellular Ca^{2+} concentrations, the Ca^{2+} ATPase pump plays a crucial role in regulating various cellular processes, including muscle contraction and neurotransmitter release.

Interplay Between Ion Channels and Pumps in Maintaining Electrical Potential

The electrical potential across cell membranes is maintained through the coordinated activity of ion channels and pumps. While ion channels regulate the passive movement of ions across the membrane in response to specific stimuli, ion pumps actively maintain the ion concentration gradients necessary for establishing the resting membrane potential and generating electrical signals. This delicate balance between the passive diffusion of ions through channels and the active transport of ions by pumps ensures the proper functioning of the electrical systems within the human body, allowing for efficient transmission of information and the regulation of various physiological processes.

The Balance Between Passive and Active Ion Transport

The interplay between ion channels and pumps is essential for maintaining the electrical potential across cell membranes and ensuring the proper functioning of cells, particularly neurons and muscle cells. To maintain this balance between passive and active ion transport, cells must continuously adjust the activity of ion channels and pumps in response to changing conditions and stimuli.

Dynamic Regulation of Ion Channels

The activity of ion channels can be dynamically regulated by various factors, such as changes in the membrane potential, ligand binding, or intracellular signalling pathways. This dynamic regulation allows cells to modulate their electrical properties in response to specific conditions, ensuring the efficient transmission of electrical signals and the precise control of cellular processes.

For example, in neurons, the opening of ligand-gated ion channels in response to neurotransmitter binding can cause depolarizing or hyperpolarizing graded potentials, which may ultimately lead to the generation of an action potential if the threshold is reached. Similarly, the activity of voltage-gated channels can be modulated by changes in the membrane potential, allowing for the rapid generation and propagation of action potentials in response to stimuli.

Adaptation of Ion Pump Activity

The activity of ion pumps can also be adjusted to maintain the appropriate ion gradients and electrical potential across cell membranes. For instance, the Na^+/K^+ ATPase pump may increase its activity in response to increased intracellular Na^+ concentrations, effectively restoring the ion gradients and the resting membrane potential.

Moreover, cells can modulate the expression and activity of ion pumps in response to long-term changes in their environment or metabolic demands. This adaptive mechanism ensures that the electrical potential across cell membranes remains stable and that the cells can continue to function efficiently under varying conditions.

Implications of Ion Channels and Pumps in Health and Disease

The proper functioning of ion channels and pumps is essential for maintaining the electrical potential across cell membranes and the overall health of the human body. Disruptions in the activity or regulation of these proteins can lead to various diseases and disorders, particularly those involving the nervous and muscular systems.

For example, mutations in ion channel genes can cause channelopathies, a group of disorders characterized by abnormal ion channel function. These conditions can manifest as epilepsy, cardiac arrhythmias, periodic paralysis, and other neurological and muscular disorders. Similarly, dysfunction of ion pumps can result in conditions such as cystic fibrosis, which is caused by mutations in the CFTR gene encoding a chloride ion channel and pump.

Understanding the role of ion channels and pumps in maintaining the electrical potential across cell membranes not only provides insights into the fundamental mechanisms underlying the electrical systems within the human body, but also offers potential therapeutic targets for the treatment of various diseases and disorders.

In conclusion, ion channels and pumps play a crucial role in maintaining the electrical potential across cell membranes by regulating the movement of ions in and out of cells. This delicate balance between passive and active ion transport ensures the proper functioning of the electrical systems within the human body, which is essential for the efficient transmission of information and the regulation of various physiological processes. Understanding these mechanisms and their implications in health and disease may pave the way for novel therapeutic approaches and improved outcomes for patients suffering from neurological and muscular disorders.

THE PROCESS OF ELECTRICAL SIGNAL TRANSMISSION ALONG NEURONS

The Process of Electrical Signal Transmission Along Neurons

The transmission of electrical signals along neurons is essential for the efficient communication of information within the nervous system. This process involves the generation and propagation of action potentials, which are rapid, transient changes in the membrane potential that travel along the neuron's axon. In this section, we will discuss the process of electrical signal transmission along neurons, focusing on the initiation, propagation, and termination of action potentials.

Initiation of Action Potentials

The initiation of an action potential in neurons typically begins with the generation of graded potentials at the dendrites, which are the neuron's input sites. Graded potentials are small, localized changes in the membrane potential that can be either depolarizing or hyperpolarizing. They are caused by the opening of ligand-gated ion channels in response to neurotransmitters binding to receptors at the synapses. The graded potentials spread passively toward the cell body and the axon hillock.

If the sum of these graded potentials is depolarizing and reaches the threshold potential (usually around -55 mV) at the axon hillock, an action potential is generated. The axon hillock has a high density of voltage-gated Na+ channels, which are crucial for initiating action potentials.

Propagation of Action Potentials

Once initiated, action potentials propagate along the axon toward the axon terminals, where they transmit the electrical signal to the next neuron or effector cell. The propagation of action potentials involves a series of well-coordinated events:

Depolarization: When the threshold potential is reached, the voltage-gated Na+ channels open, causing a rapid influx of Na+ ions and a further depolarization of the membrane. This local depolarization, in turn, causes adjacent voltage-gated Na+ channels to open, allowing the influx of more Na+ ions and the propagation of the action potential along the axon.

Repolarization: As the membrane potential approaches +30 mV, the voltage-gated Na+ channels begin to inactivate, while voltage-gated K+ channels open, allowing K+ ions to flow out of the cell. This efflux of K+ ions repolarize the membrane, restoring the negative membrane potential.

Refractory periods: Neurons exhibit refractory periods, during which they are either incapable (absolute refractory period) or less likely (relative refractory period) to generate another action potential. These refractory periods ensure unidirectional propagation of action potentials along the axon and prevent continuous firing of action potentials.

The speed of action potential propagation is influenced by factors such as axon diameter and the presence of myelin sheaths. Larger axon diameters and myelinated axons allow for faster propagation of action potentials due to decreased resistance and increased insulation, respectively.

Termination of Action Potentials and Signal Transmission

Action potentials propagate along the axon until they reach the axon terminals, which are specialized structures that make contact with the next neuron or effector cell at the synapse. At the synapse, the electrical signal is typically converted into a chemical signal through the release of neurotransmitters. This process is known as synaptic transmission and involves the following steps:

Arrival of the action potential: When the action potential reaches the axon terminal, it causes the opening of voltage-gated Ca2+ channels, allowing Ca2+ ions to enter the terminal.

Neurotransmitter release: The influx of Ca^{2+} ions trigger the fusion of neurotransmitter-containing vesicles with the presynaptic membrane, releasing neurotransmitters into the synaptic cleft.

Postsynaptic signalling: The neurotransmitters diffuse across the synaptic cleft and bind to specific receptors on the postsynaptic membrane of the next neuron or effector cell. This binding causes the opening of ligand-gated ion channels on the postsynaptic membrane, leading to the generation of graded potentials in the postsynaptic cell.

Signal integration: The graded potentials generated in the postsynaptic cell can be either excitatory or inhibitory, depending on the type of neurotransmitter and receptor involved. If the sum of these graded potentials reaches the threshold potential at the axon hillock of the postsynaptic neuron, an action potential is generated, and the electrical signal is transmitted to the next neuron in the chain.

Termination of neurotransmitter action: The action of neurotransmitters in the synaptic cleft is terminated through various mechanisms, such as reuptake into the presynaptic neuron, diffusion away from the synapse, or enzymatic degradation. This termination ensures that the signal transmission is temporally precise and prevents continuous activation of the postsynaptic cell.

Modulation of Neuronal Signal Transmission

The transmission of electrical signals along neurons and the propagation of action potentials can be modulated by various factors, including the intrinsic properties of the neurons, the presence of inhibitory or excitatory inputs, and the activity of neuromodulators.

Intrinsic properties: The intrinsic properties of neurons, such as the density and distribution of ion channels and the morphology of the axon, can influence the initiation and propagation of action potentials. For example, neurons with a higher density of voltage-gated Na+ channels at the axon hillock may have a lower threshold for action potential generation, while neurons with larger axon diameters or myelinated axons may exhibit faster action potential propagation.

Inhibitory and excitatory inputs: The balance between inhibitory and excitatory inputs can modulate the likelihood of action potential generation in a neuron. Inhibitory inputs typically hyperpolarize the membrane potential, making it less likely for the neuron to reach the threshold potential, while excitatory inputs depolarize the membrane potential, increasing the likelihood of action potential generation.

Neuromodulators: Neuromodulators are signalling molecules that can modulate the transmission of electrical signals along neurons by altering the activity of ion channels, neurotransmitter release, or synaptic transmission. Some common neuromodulators include dopamine, serotonin, and acetylcholine, which can exert their effects through the activation of various receptor types and intracellular signalling pathways.

In conclusion, the process of electrical signal transmission along neurons involves the initiation, propagation, and termination of action potentials, which are crucial for the efficient communication of information within the nervous system. The propagation of action potentials is influenced by factors such as axon diameter, myelination, and the balance between inhibitory and excitatory inputs. Understanding these mechanisms and their modulation by various factors is essential for gaining insights into the complex electrical systems within the human body and their roles in diverse physiological processes.

THE INFLUENCE OF ELECTRICAL SYSTEMS ON SENSORY PERCEPTION

Electrical systems within the human body play a crucial role in sensory perception, which is the process by which our nervous system detects, processes, and interprets information from the external environment. Sensory receptors, specialized neurons or cells, convert various types of energetic stimuli (e.g., light, sound, temperature, pressure) into electrical signals that can be processed by the nervous system.

Sensory Transduction and Signal Transmission

The process of sensory transduction involves converting external stimuli into electrical signals through the activation of specific sensory receptors. For example, photoreceptors in the retina convert light into electrical signals, while mechanoreceptors in the skin respond to pressure or vibration. When sensory receptors are activated, they generate graded potentials that, if strong enough, can lead to the generation of action potentials in sensory neurons.

Sensory neurons transmit these electrical signals through their axons to the central nervous system, where they are processed and integrated by specialized brain regions. This signal transmission and processing allow us to perceive and interpret various aspects of our environment, such as vision, hearing, touch, taste, and smell.

The Influence of Electrical Systems on Muscle Contraction

Electrical systems are also essential for the regulation of muscle contraction, which is the process by which muscles generate force and movement. Muscle contraction is initiated by the generation of electrical signals in motor neurons, which transmit these signals to muscle cells through neuromuscular junctions.

Excitation-Contraction Coupling

The process of excitation-contraction coupling involves the conversion of electrical signals in motor neurons into mechanical force in muscle cells. When an action potential reaches the axon terminal of a motor neuron, it triggers the release of the neurotransmitter acetylcholine, which binds to receptors on the muscle cell membrane. This binding causes the opening of ligand-gated ion channels, leading to a depolarization of the muscle cell membrane and the generation of an action potential in the muscle cell.

The action potential in the muscle cell propagates along the cell membrane and into the interior of the cell through specialized structures called T-tubules. This propagation causes the release of Ca2+ ions from the sarcoplasmic reticulum, a specialized intracellular compartment that stores Ca2+ ions. The increase in intracellular Ca2+ ions trigger the interaction between actin and myosin filaments within the muscle cell, leading to muscle contraction.

Muscle Relaxation

Muscle relaxation occurs when the electrical signal in the muscle cell ceases, and intracellular Ca2+ ion concentrations decrease. The decrease in Ca2+ ions is facilitated by the Ca2+ ATPase pump, which actively transports Ca2+ ions back into the sarcoplasmic reticulum. The reduction in Ca2+ ions prevent the interaction between actin and myosin filaments, allowing the muscle cell to return to its resting state.

The Influence of Electrical Systems on Overall Body Function

The electrical systems within the human body are essential for the regulation and coordination of various physiological processes, including sensory perception, muscle contraction, and overall body function. For example, electrical signals generated in response to sensory stimuli can influence motor output, leading to the appropriate adjustments in muscle activity and movement. Additionally, electrical signals are critical for the regulation of essential processes, such as heart rate, respiration, and digestion.

Moreover, the electrical activity in the nervous system underlies complex cognitive functions, including learning, memory, and decision-making. Neuronal networks in the brain generate and transmit electrical signals that are responsible for processing and integrating information, allowing us to adapt our behaviour and responses to the environment.

In conclusion, electrical systems within the human body have a profound impact on sensory perception, muscle contraction, and overall function. They enable us to perceive and interpret our surroundings, generate coordinated movements, and regulate various physiological processes essential for survival. These electrical systems also contribute to our cognitive functions, such as learning, memory, and decision-making, highlighting their importance in shaping human behaviour and experience.

THE INTEGRATION OF ELECTRICAL SYSTEMS FOR COORDINATED FUNCTION

The coordination and integration of electrical systems within the human body are vital for maintaining homeostasis and ensuring proper functioning. For instance, the autonomic nervous system, a division of the peripheral nervous system, is responsible for regulating involuntary processes, such as heart rate, digestion, and respiration. The autonomic nervous system is further divided into the sympathetic and parasympathetic divisions, which often have opposing effects on target organs and work together to maintain homeostasis.

The central nervous system, consisting of the brain and spinal cord, plays a crucial role in integrating sensory information and coordinating appropriate motor responses. Sensory information from peripheral receptors is processed and interpreted by the brain, which generates motor commands that are transmitted to the muscles via motor neurons. This integrated system ensures that our body responds appropriately to various environmental stimuli and internal physiological changes.

In addition, electrical systems are critical for maintaining communication between different body systems. For example, the endocrine system, which regulates hormones, interacts with the nervous system through neuroendocrine signalling. This interaction allows for the fine-tuning of physiological responses, such as the stress response, growth, and reproduction, based on the integration of hormonal and neuronal signals.

Overall, the electrical systems within the human body serve as a foundation for the complex interactions between various physiological processes and the coordination of sensory perception, muscle contraction, and overall function. Understanding these systems and their integration is essential for gaining insights into human health and disease, as well as developing potential therapeutic interventions for disorders involving the nervous system or muscle function.

The Potential Applications of Electrical System Research in Medical Technologies

Research into the electrical systems within the human body has paved the way for the development of innovative medical technologies that can significantly enhance the quality of life for individuals with various neurological or muscular disorders. Two prominent examples of these technologies are neuroprosthetics and brain-computer interfaces (BCIs), which directly interface with the nervous system to restore or augment function. In this section, we will discuss the potential applications of electrical system research in the development of these medical technologies and their potential impact on healthcare.

Neuroprosthetics

Neuroprosthetics are medical devices designed to replace or supplement the function of a damaged or lost body part by interfacing directly with the nervous system. These devices can be used to restore sensory perception, motor function, or cognitive abilities in individuals affected by neurological or muscular disorders, such as amputations, spinal cord injuries, or strokes.

Sensory Neuroprosthetics

Sensory neuroprosthetics aim to restore sensory perception by converting external stimuli into electrical signals that can be processed by the nervous system. For example, cochlear implants can restore hearing in individuals with profound hearing loss by converting sound waves into electrical signals that directly stimulate the auditory nerve. Similarly, retinal implants have been developed to restore vision in individuals with retinal degeneration by converting light into electrical signals that stimulate the retinal ganglion cells. Specifically, new technologies are being developed to alter sensory challenged individuals, as well as neuroprosthetics to help alter epilepsy and OCD.

Motor Neuroprosthetics

Motor neuroprosthetics are designed to restore motor function by detecting and decoding the electrical signals generated by the nervous system during the planning or execution of movement. These devices can then use this information to control external prosthetic limbs or stimulate paralyzed muscles. For example, myoelectric prosthetic limbs can detect electrical signals from the residual muscles in an amputated limb and use this information to control the movement of the prosthetic limb. Similarly, functional electrical stimulation (FES) can be used to stimulate paralyzed muscles, enabling individuals with spinal cord injuries or other neurological disorders to regain some degree of motor function.

Brain-Computer Interfaces (BCIs)

Brain-computer interfaces (BCIs) are systems that enable direct communication between the brain and external devices by detecting and decoding the electrical signals generated by the brain. BCIs have the potential to revolutionize the treatment and rehabilitation of individuals with various neurological or muscular disorders by allowing them to control computers, communication devices, or even robotic limbs using their thoughts.

Non-invasive BCIs

Non-invasive BCIs typically rely on electroencephalography (EEG) to detect the electrical activity of the brain from the scalp. These systems can be used for a variety of applications, such as enabling individuals with severe motor disabilities, like those suffering from locked-in syndrome or amyotrophic lateral sclerosis (ALS), to communicate or control a computer cursor using their brain activity.

Invasive BCIs

Invasive BCIs involve the implantation of electrodes directly into the brain to detect neural activity with higher spatial resolution and signal-to-noise ratio than non-invasive BCIs. These systems have shown promise in restoring motor function in individuals with paralysis by decoding the neural signals associated with the intention to move and using this information to control robotic limbs or FES systems.

In conclusion, research into the electrical systems within the human body has the potential to significantly impact the development of innovative medical technologies, such as neuroprosthetics and brain-computer interfaces. These technologies can restore or augment sensory perception, motor function, and cognitive abilities in individuals affected by various neurological or muscular disorders, ultimately improving their quality of life and expanding the possibilities for treatment and rehabilitation. Continued research in this area is essential for advancing our understanding of the human body's electrical systems and for refining and expanding the applications of these medical technologies.

FUTURE DIRECTIONS AND CHALLENGES

As the understanding of electrical systems within the human body continues to grow, there is potential for the development of new and more advanced neuroprosthetic and BCI technologies. Some future directions and challenges in this field include:

Improving the biocompatibility and long-term stability of implanted devices: Ensuring that implanted neuroprosthetic devices and BCIs remain functional over extended periods and do not cause adverse tissue reactions is crucial for the widespread adoption of these technologies.

Enhancing the resolution and accuracy of neural signal detection and decoding: Improving the ability to detect and interpret the electrical signals generated by the nervous system will enable more precise control of neuroprosthetic devices and BCIs, leading to better functional outcomes for paticnts.

Developing closed-loop systems: Integrating sensory feedback into neuroprosthetic and BCI systems will enable users to better control these devices and improve the overall performance of the system. For example, providing tactile feedback in prosthetic limbs could help users better grasp objects and manipulate their environment.

Expanding the range of applications: As our understanding of the electrical systems within the human body continues to grow, there is potential for developing new neuroprosthetic and BCI applications, such as devices that restore or augment cognitive function in individuals with brain injuries or neurodegenerative disorders.

Addressing ethical and social considerations: The development and widespread adoption of neuroprosthetic devices and BCIs raise various ethical and social concerns, such as issues related to privacy, autonomy, and the potential for enhancing human capabilities beyond their natural limits. Addressing these concerns will be essential for ensuring the responsible development and implementation of these technologies.

Overall, research into the electrical systems within the human body holds significant promise for the development of innovative medical technologies, such as neuroprosthetics and brain-computer interfaces. These technologies have the potential to improve the quality of life for individuals with neurological or muscular disorders by restoring or augmenting function and expanding the possibilities for treatment and rehabilitation. As our understanding of these systems continues to grow, so too will the potential for new and ground-breaking applications that can revolutionize healthcare and positively impact the lives of countless individuals.

Biophysics and Electrophysiology for Studying Electrical Signal Generation and Propagation

Biophysics and electrophysiology are interdisciplinary fields that combine principles from physics, biology, and engineering to study the electrical properties of biological systems. These approaches can provide valuable insights into the fundamental principles of electrical signal generation and propagation in cells, particularly neurons and muscle cells.

Techniques in Electrophysiology

Electrophysiology employs a variety of experimental techniques to measure and manipulate the electrical properties of cells, tissues, and organisms. Some key techniques include:

Patch-clamp: This technique allows for the direct measurement of ion channel activity in cells by recording the flow of ions across the cell membrane. Patch-clamp can be used to study the properties of individual ion channels, as well as to investigate how these channels are modulated by various factors, such as voltage, ligands, or intracellular signalling pathways.

Voltage and current clamp: These techniques involve the controlled manipulation of membrane voltage or current to study the electrical properties of cells, such as membrane resistance, capacitance, and the generation of action potentials. These methods can be used to investigate the mechanisms underlying electrical signal generation and propagation in neurons and muscle cells.

Multi-electrode arrays (MEAs): MEAs enable the simultaneous recording of electrical activity from multiple neurons or cell populations in vitro or in vivo. This technique can be used to study the network-level properties of neuronal circuits and how they process and transmit information.

Molecular Biology and Biochemistry for Examining Ion Channels and Pumps

Molecular biology and biochemistry are essential for understanding the structure and function of ion channels, pumps, and other key components involved in maintaining the body's electrical systems. These approaches can provide detailed information about the molecular mechanisms that govern ion transport and electrical signal generation.

Techniques in Molecular Biology and Biochemistry

Some key techniques in molecular biology and biochemistry that can be used to study ion channels, pumps, and related components include:

Gene cloning and expression: Molecular cloning techniques enable researchers to isolate and amplify specific genes that encode ion channels or pumps. These genes can then be expressed in heterologous systems, such as cultured cells or Xenopus oocytes, to study the functional properties of the proteins they encode.

Protein purification and reconstitution: Biochemical methods can be used to purify ion channels or pumps from native tissues or heterologous expression systems. The purified proteins can then be reconstituted into artificial lipid membranes or proteoliposomes to study their functional properties in a controlled environment.

X-ray crystallography and cryo-electron microscopy: These structural biology techniques allow for the determination of high-resolution structures of ion channels and pumps. Understanding the three-dimensional structure of these proteins can provide critical insights into their function, such as the mechanisms of ion selectivity and gating.

Site-directed mutagenesis: This technique involves the targeted modification of specific amino acid residues in ion channels or pumps to investigate their functional roles. By introducing point mutations or deletions, researchers can probe the importance of specific residues or domains for ion transport and gating.

By combining biophysical and electrophysiological approaches with molecular biology and biochemistry, researchers can gain a comprehensive understanding of the fundamental principles governing electrical signal generation and propagation in biological systems. This knowledge can not only enhance our understanding of the body's electrical systems but also contribute to the development of novel therapies and medical technologies for disorders related to these systems.

Biophysics and Electrophysiology for Studying Electrical Signal Generation and Propagation in Biological Systems

As previously mentioned, biophysics and electrophysiology combine principles from physics, biology, and engineering to study the electrical properties of biological systems. These approaches provide valuable insights into the fundamental principles of electrical signal generation and propagation in cells, particularly neurons and muscle cells, and are essential for understanding the specific processes and interactions of electrical signals in the nervous system.

Neuroscience and Neurophysiology: Understanding Electrical Signals in the Nervous System

Neuroscience is an interdisciplinary field that seeks to understand the structure, function, and processes of the nervous system. Neurophysiology, a sub-discipline of neuroscience, specifically focuses on the physiological processes and interactions of neurons and neuronal networks, including the generation and transmission of electrical signals. By integrating knowledge from biophysics and electrophysiology with neuroscience and neurophysiology, researchers can develop a comprehensive understanding of the electrical signalling processes within the nervous system.

Techniques in Neuroscience and Neurophysiology

Some key techniques used in neuroscience and neurophysiology to study electrical signals in the nervous system include:

Intracellular and extracellular recordings: Intracellular recordings involve the insertion of an electrode into a single neuron to measure its electrical properties and activity, while extracellular recordings involve placing an electrode near a neuron or a group of neurons to measure their collective activity. These techniques can provide insights into the mechanisms underlying electrical signal generation, synaptic transmission, and network-level processing in the nervous system.

Optogenetics: Optogenetics involves the genetic modification of neurons to express light-sensitive ion channels or receptors, allowing researchers to control neuronal activity using light. This powerful technique can be used to investigate the role of specific neurons or neuronal circuits in various behaviours and cognitive processes.

Calcium imaging: Calcium imaging techniques involve the use of fluorescent calcium indicators to visualize changes in intracellular calcium levels, which are often associated with neuronal activity. Calcium imaging can be used to monitor the activity of individual neurons or large populations of neurons simultaneously, providing insights into the spatiotemporal dynamics of electrical signalling in the nervous system.

Functional magnetic resonance imaging (fMRI) and magnetoencephalography (MEG): These non-invasive imaging techniques enable researchers to study the activity of the human brain in vivo. By measuring changes in blood flow (fMRI) or magnetic fields (MEG) related to neuronal activity, researchers can investigate the functional organization of the brain and the neural basis of various cognitive processes.

Applications in Neuroscience and Neurophysiology

The integration of biophysics, electrophysiology, neuroscience, and neurophysiology has led to numerous applications and advances in our understanding of the nervous system. Some key applications include:

Identifying the cellular and molecular mechanisms underlying various neurological disorders, such as epilepsy, Parkinson's disease, and Alzheimer's disease.

Developing targeted therapies for neurological disorders based on the modulation of specific ion channels, neurotransmitter systems, or neuronal circuits.

Investigating the neural basis of perception, cognition, and behaviour, as well as the changes that occur during learning, memory formation, and neuroplasticity.

Designing neural interfaces and prosthetic devices that can restore or augment sensory perception, motor function, or cognitive abilities in individuals with neurological or muscular disorders.

In summary, the integration of biophysics, electrophysiology, neuroscience, and neurophysiology enables researchers to study the fundamental principles of electrical signal generation and propagation in biological systems and understand the specific processes and interactions of electrical signals in the nervous system. This comprehensive understanding is crucial for advancing our knowledge of the nervous system's structure and function, as well as for developing novel therapies and medical technologies for disorders related to these systems.

Future Perspectives and Challenges

As our understanding of the electrical signalling processes in the nervous system continues to grow, new opportunities and challenges will emerge in the field. Some future perspectives and challenges include:

Developing a more detailed understanding of the interactions between electrical signalling and other cellular processes, such as gene expression, protein synthesis, and metabolic regulation.

Investigating the role of electrical signalling in non-neuronal cells, such as glial cells, and understanding how these cells contribute to the overall function of the nervous system.

Exploring the role of electrical signalling in the development and maturation of the nervous system, as well as in the processes of aging and neurodegeneration.

Addressing the ethical and societal implications of advances in our understanding of electrical signalling in the nervous system, particularly as they relate to the development of novel therapies, medical technologies, and brain-computer interfaces.

In conclusion, the integration of biophysics, electrophysiology, neuroscience, and neurophysiology is essential for understanding the fundamental principles of electrical signal generation and propagation in biological systems, and for examining the specific processes and interactions of electrical signals in the nervous system. This multidisciplinary approach will continue to drive advances in our understanding of the nervous system, as well as the development of novel therapies and medical technologies for disorders related to these systems.

BIOPHYSICS AND ELECTROPHYSIOLOGY FOR STUDYING ELECTRICAL SIGNAL GENERATION AND PROPAGATION IN BIOLOGICAL SYSTEMS

As previously discussed, biophysics and electrophysiology are interdisciplinary fields that utilize principles from physics, biology, and engineering to study the electrical properties of biological systems. These approaches are crucial for understanding the fundamental principles of electrical signal generation and propagation in cells, particularly neurons and muscle cells. This knowledge serves as the foundation for clinical research and case studies exploring the implications of electrical systems in medical diagnosis and treatment.

Clinical Research and Case Studies: Exploring the Implications of Electrical Systems in Medical Diagnosis and Treatment

Clinical research and case studies are valuable tools for investigating the practical applications and implications of electrical systems in medicine. By examining the real-world experiences of patients and healthcare providers, these approaches can provide insights into the diagnosis and treatment of various medical conditions related to electrical systems in the human body.

Applications of Clinical Research and Case Studies

Some key applications of clinical research and case studies in the context of electrical systems in medicine include:

Investigating the effectiveness of novel diagnostic tools and therapies: Clinical research can help evaluate the efficacy and safety of new diagnostic methods and treatments targeting the electrical systems in the human body. For example, clinical trials can be conducted to test the effectiveness of drugs that modulate ion channels or neurotransmitter systems in treating various neurological disorders.

Identifying factors that influence patient outcomes: Case studies can provide insights into the factors that contribute to the success or failure of specific diagnostic methods or treatments in individual patients. This information can help inform personalized medicine approaches and the development of tailored treatment plans based on a patient's unique characteristics.

Understanding the long-term effects and potential complications of treatments: Clinical research and case studies can help identify the long-term effects and potential complications of treatments targeting the electrical systems in the human body, such as the use of deep brain stimulation for Parkinson's disease or the implantation of cardiac pacemakers for heart rhythm disorders.

Evaluating the real-world effectiveness of medical devices and technologies: Clinical research and case studies can provide valuable information about the real-world effectiveness of medical devices and technologies that interface with the electrical systems in the human body, such as cochlear implants, retinal prostheses, or brain-computer interfaces.

Challenges and Future Directions in Clinical Research and Case Studies

Some challenges and future directions in clinical research and case studies related to electrical systems in medicine include:

Developing more effective diagnostic tools and therapies: As our understanding of the electrical systems in the human body continues to grow, there is potential for the development of more accurate and sensitive diagnostic tools, as well as novel therapies targeting specific ion channels, neurotransmitter systems, or neuronal circuits.

Expanding access to cutting-edge treatments and technologies: Ensuring that patients have access to the latest diagnostic methods, treatments, and medical technologies related to electrical systems in the human body is a critical challenge. Efforts to reduce disparities in access to care and promote the widespread adoption of these innovations will be essential for improving patient outcomes.

Addressing ethical and social considerations: The development and application of novel diagnostic methods, treatments, and medical technologies related to electrical systems in the human body raise various ethical and social concerns. These concerns may include issues related to patient privacy, autonomy, and the potential for enhancing human capabilities beyond their natural limits. Addressing these concerns will be crucial for ensuring the responsible and equitable development and implementation of these innovations.

In summary, clinical research and case studies play a critical role in exploring the implications of electrical systems in medical diagnosis and treatment. By integrating knowledge from biophysics and electrophysiology with clinical research and case studies, researchers and healthcare providers can develop a deeper understanding of the practical applications and potential limitations of these systems in medicine.

Here are some case studies related to electrical systems in the human body and their implications for medical diagnosis and treatment:

1. Deep Brain Stimulation (DBS) for Parkinson's Disease Case study: Volkmann J, et al. (2006). Deep brain stimulation for the treatment of Parkinson's disease: subthalamic nucleus versus globus pallidus internus. Journal of Neurology, Neurosurgery, and Psychiatry, 77(4), 464-470.

This case study compares the efficacy of deep brain

stimulation (DBS) targeting the subthalamic nucleus (STN) and globus pallidus internus (GPi) in patients with Parkinson's disease. The study found that both targets provided significant improvements in motor function, but STN stimulation resulted in greater improvements in medication reduction.

2. Transcranial Magnetic Stimulation (TMS) for Depression Case study: O'Reardon JP, et al. (2007). Efficacy and safety of transcranial magnetic stimulation in the acute treatment of major depression: a multisite randomized controlled trial. Biological Psychiatry, 62(11), 1208-1216.

This case study investigated the efficacy and safety of repetitive transcranial magnetic stimulation (rTMS) for the acute treatment of major depressive disorder. The results showed that rTMS was effective in reducing depressive symptoms and was well-tolerated by patients.

3. Cochlear Implants for Sensorineural Hearing Loss Case study: Gifford RH, et al. (2013). Cochlear implantation with hearing preservation yields significant benefit for speech recognition in complex listening environments. Ear and Hearing, 34(4), 413-425.

This case study examined the benefits of cochlear implantation with hearing preservation in patients with sensorineural hearing loss. The results indicated that this approach significantly improved speech recognition in complex listening environments compared to preoperative performance with hearing aids.

4. Implantable Cardioverter-Defibrillators (ICDs) for Ventricular Arrhythmias Case study: Moss AJ, et al. (2002). Prophylactic implantation of a defibrillator in patients with myocardial infarction and reduced ejection fraction. The New England Journal of Medicine, 346(12), 877-883.

This case study investigated the effectiveness of prophylactic implantable cardioverter-defibrillator (ICD) therapy in patients with a history of myocardial infarction and reduced ejection fraction. The study found that ICD therapy significantly reduced the risk of death from arrhythmias and improved overall survival.

5. Brain-Computer Interfaces (BCIs) for Motor Disabilities Case study: Hochberg LR, et al. (2012). Reach and grasp by people with tetraplegia using a neurally controlled robotic arm. Nature, 485(7398), 372-375.

This case study demonstrated the use of a brain-computer interface (BCI) to control a robotic arm in patients with tetraplegia due to spinal cord injury. The results showed that patients were able to successfully reach and grasp objects using the BCI-controlled robotic arm, highlighting the potential of BCIs to restore motor function in individuals with severe motor disabilities.

These case studies illustrate the potential of electrical systems in the human body for diagnosing and treating various medical conditions. The advances in our understanding of these systems have led to the development of novel therapies and medical technologies that have the potential to improve the lives of countless patients.

Biophysics and Electrophysiology: Studying Electrical Signal Generation and Propagation in Biological Systems

Biophysics and electrophysiology are interdisciplinary fields that apply the principles of physics, biology, and engineering to study the electrical properties of biological systems. These approaches are essential for understanding the fundamental principles of electrical signal generation and propagation in cells, particularly neurons and muscle cells. This knowledge lays the foundation for the development of innovative medical technologies by integrating technology and engineering to identify and evaluate potential applications of electrical system research.

Technology and Engineering: Identifying and Evaluating Potential Applications of Electrical System Research in the Development of Innovative Medical Technologies

The combination of technology and engineering with the knowledge gained from biophysics and electrophysiology enables the development of cutting-edge medical technologies. These technologies have the potential to revolutionize healthcare by improving diagnostics, treatments, and the overall quality of life for patients with various medical conditions related to electrical systems in the human body.

Applications of Technology and Engineering in the Development of Medical Technologies

Some key applications of technology and engineering in the development of innovative medical technologies include:

Designing and optimizing medical devices: Engineers can use their expertise in materials science, electronics, and mechanical engineering to design and optimize medical devices that interact with electrical systems in the human body, such as cardiac pacemakers, cochlear implants, and brain-computer interfaces.

Developing advanced neuroimaging techniques: Technological advances in imaging systems, data processing, and computational modelling can lead to the development of more sensitive and accurate neuroimaging techniques that allow researchers and clinicians to study brain function and dysfunction with greater precision.

Creating wearable and implantable sensors: The integration of advanced materials, miniaturized electronics, and wireless communication technologies enables the development of wearable and implantable sensors that can monitor various physiological parameters related to electrical systems in the human body, such as heart rate, muscle activity, and neural activity.

Engineering tissue and cellular models: Advances in tissue engineering, microfabrication, and microfluidics can lead to the development of engineered tissue and cellular models that can be used to study the electrical properties of cells and tissues in a controlled and reproducible environment, providing valuable insights into the mechanisms underlying electrical signal generation and propagation.

Challenges and Future Directions in Technology and Engineering for Medical Technologies

Some challenges and future directions in technology and engineering for the development of innovative medical technologies related to electrical systems in the human body include:

Developing more effective and targeted therapies: As our understanding of the electrical systems in the human body continues to grow, there is potential for the development of more effective and targeted therapies, such as drugs that modulate ion channels or neurotransmitter systems, or electrical stimulation devices that target specific neuronal circuits.

Ensuring safety and biocompatibility: It is crucial to ensure that medical devices and technologies designed to interface with the electrical systems in the human body are safe and biocompatible, minimizing the risk of adverse reactions or complications in patients.

Addressing ethical and social considerations: The development and application of novel medical technologies related to electrical systems in the human body raise various ethical and social concerns, such as issues related to patient privacy, autonomy, and the potential for enhancing human capabilities beyond their natural limits. Addressing these concerns will be crucial for ensuring the responsible and equitable development and implementation of these innovations.

In summary, the integration of biophysics, electrophysiology, technology, and engineering is essential for identifying and evaluating potential applications of electrical system research in the development of innovative medical technologies. This interdisciplinary approach will continue to drive advances in the understanding of electrical systems in the human body and contribute to the development of novel therapies and medical technologies for disorders related to these systems.

BONES

Although bones may not be typically associated with electrical systems, they do possess electrical properties that play an essential role in their function and health. One of the key electrical phenomena in bones is piezoelectricity, which is the generation of an electrical charge in response to mechanical stress. Additionally, bones exhibit a property called streaming potential, which is the generation of electrical signals due to fluid movement within the bone's porous structure. These electrical properties are crucial for understanding bone physiology, remodelling, and repair processes.

Piezoelectricity in Bones

Piezoelectricity is a property exhibited by certain materials, including bones, that generates an electrical charge in response to mechanical stress. In bones, this electrical property is primarily attributed to the collagen fibrils and hydroxyapatite crystals, which are the primary structural components of the bone matrix. When a bone is subjected to mechanical stress, such as during movement or loading, the collagen fibrils and hydroxyapatite crystals become distorted, generating an electrical charge.

The piezoelectric properties of bones have several important implications:

Bone remodelling: The electrical charges generated by piezoelectricity may play a role in the bone remodelling process by signalling bone-forming cells (osteoblasts) and bone-resorbing cells (osteoclasts). The electrical signals produced by mechanical stress may help regulate the balance between bone formation and resorption, contributing to the maintenance of healthy bone structure.

Bone repair: The electrical charges generated by piezoelectricity may also have a role in bone repair and fracture healing. Studies have shown that the application of electrical stimulation can promote bone repair by enhancing the recruitment and differentiation of bone-forming cells at the fracture site.

Streaming Potential in Bones

Streaming potential is another electrical property of bones, which is the generation of electrical signals due to fluid movement within the porous structure of the bone. Bones contain a network of interconnected pores filled with interstitial fluid. When mechanical stress is applied to the bone, the fluid within these pores is forced to move, generating electrical signals.

The streaming potential has several important implications for bone physiology:

Mechanotransduction: Streaming potential is thought to be one of the mechanisms by which bone cells sense and respond to mechanical loading. The electrical signals generated by fluid movement within the bone may activate mechanosensitive ion channels on the surface of bone cells, leading to changes in intracellular signalling pathways and gene expression.

Bone remodelling and repair: Similar to piezoelectricity, the electrical signals generated by streaming potential may play a role in regulating the activity of bone-forming and bone-resorbing cells, thus contributing to bone remodelling and repair processes.

In conclusion, bones possess electrical properties such as piezoelectricity and streaming potential that are crucial for understanding their function and health. These electrical properties may play a role in bone remodelling, mechanotransduction, and repair processes, and further research into these phenomena could contribute to the development of novel therapeutic strategies for bone-related disorders.

Piezoelectricity, the generation of an electrical charge in response to mechanical stress, has significant potential for medical advancement. The piezoelectric properties of biological materials, particularly bones and tendons, have led to various applications in diagnostics, treatment, and the development of novel medical devices. The following are some possible routes for medical advancement utilizing piezoelectricity:

Bone fracture healing and repair: The piezoelectric properties of bones have been found to play a role in bone repair and fracture healing. Electrical stimulation, such as pulsed electromagnetic fields (PEMF) or direct current stimulation, has been shown to promote bone repair by enhancing the recruitment, differentiation, and activity of bone-forming cells (osteoblasts) at the fracture site. Developing devices that exploit the piezoelectric effect in bones could provide non-invasive and targeted treatments for bone fractures and other bone-related disorders.

Tissue engineering and regenerative medicine: Piezoelectric materials can be incorporated into tissue engineering scaffolds to mimic the native piezoelectric properties of biological tissues. These scaffolds can be used to create an electroactive environment that promotes cell attachment, proliferation, and differentiation, which is essential for tissue regeneration. For example, piezoelectric scaffolds have been used in the regeneration of bone, cartilage, and neural tissues.

Energy harvesting for implantable devices: The piezoelectric effect can be used to harvest energy from the mechanical movements of the body, such as breathing, heartbeat, or walking. This energy can be used to power implantable medical devices, such as pacemakers, sensors, and drug delivery systems. By eliminating the need for batteries or external power sources, the development of self-powered implantable devices could significantly reduce the risk of infection, device failure, and the need for surgical replacement.

Biosensors and diagnostics: Piezoelectric materials can be used to develop highly sensitive biosensors for the detection of various biomolecules and physiological parameters. For example, piezoelectric sensors can be used to measure pressure, strain, or vibration in tissues, which can provide valuable information for the diagnosis and monitoring of various medical conditions, such as osteoporosis, arthritis, and muscle disorders.

Therapeutic ultrasound: Piezoelectric materials can convert electrical energy into mechanical vibrations, which can be used to generate focused ultrasound waves for therapeutic purposes. Therapeutic ultrasound has been used in various medical applications, such as promoting tissue healing, reducing inflammation, and breaking up kidney stones or tumour tissue.

Neurostimulation: Piezoelectric materials can be used to develop non-invasive or minimally invasive devices for neurostimulation, such as transcranial magnetic stimulation (TMS) or peripheral nerve stimulation. These devices could provide targeted treatment for various neurological disorders, such as depression, epilepsy, and chronic pain, without the need for invasive surgical procedures or the systemic side effects of medications.

In summary, piezoelectricity offers several promising routes for medical advancement, including bone fracture healing, tissue engineering, energy harvesting for implantable devices, biosensors, therapeutic ultrasound, and neurostimulation. Continued research and development in these areas have the potential to revolutionize healthcare by providing more effective, targeted, and minimally invasive treatments for a wide range of medical conditions.

SUMMARY

Throughout this investigation, we have emphasized the importance of interdisciplinary research, incorporating biophysics, electrophysiology, molecular biology, biochemistry, neuroscience, neurophysiology, clinical research, and technology and engineering to study the electrical systems in the human body. This multifaceted approach enables the identification and evaluation of potential applications for electrical system research in the development of novel medical technologies.

In conclusion, this comprehensive analysis of the electrical systems within the human body not only contributes to our current understanding of the fundamental principles governing these systems and their role in maintaining overall health and function but also holds the potential to revolutionize medical advancements. By exploring novel medical applications and technologies, we can pave the way for improved diagnostic and therapeutic options for patients suffering from various neurological and muscular disorders, as well as other conditions related to the body's electrical systems. This investigation serves as a foundation for future research and innovation, ultimately benefiting healthcare and the lives of patients worldwide.

A deeper understanding of the inner electrical systems of the human body can inspire the development of innovative technologies and solutions in various sectors of society. By mimicking the biological mechanisms underlying these electrical systems, we can harness their unique properties and efficiencies to address real-world challenges. Here are some examples of how this knowledge can be applied on a larger scale for various uses within society:

Bio-inspired robotics and artificial intelligence: Studying the electrical systems of the human body, particularly the nervous system, can help us develop more advanced and efficient robotic systems and artificial intelligence. By mimicking the neural networks and the way they process and transmit information, we can create bio-inspired algorithms and control systems that can enable robots to adapt to their environment, learn from experience, and perform complex tasks more efficiently.

Energy generation and storage: The human body's electrical systems, such as the process of energy conversion in cellular respiration, can inspire new ways to generate and store energy on a larger scale. For example, bio-inspired fuel cells can harness the principles of biological energy conversion to generate electricity more efficiently and sustainably. Additionally, studying the electrical properties of biological materials, like piezoelectricity in bones, can lead to the development of innovative energy harvesting technologies that convert mechanical energy into electrical energy.

Smart materials and sensors: Understanding the electrical properties of biological materials can lead to the development of smart materials and sensors that respond to changes in their environment. For example, materials that exhibit piezoelectric properties can be used to create sensors that detect pressure, strain, or vibration. These bio-inspired sensors can be applied in various industries, such as healthcare, automotive, aerospace, and infrastructure monitoring.

Communication and networking: The study of electrical systems in the human body, particularly the way neurons communicate, can inspire new strategies for communication and networking in technological systems. By mimicking the efficiency and adaptability of biological communication networks, we can develop more robust, resilient, and energy-efficient communication systems for various applications, such as the Internet of Things (IoT) and wireless sensor networks.

Environmental monitoring and remediation: Understanding the electrical systems of the human body can also inspire new approaches to environmental monitoring and remediation. For example, bio-inspired sensors that mimic the electrical properties of biological materials can be used to detect and monitor various environmental parameters, such as temperature, humidity, and air quality. Moreover, studying the electrical properties of microorganisms can lead to the development of innovative bioremediation techniques that utilize their unique abilities to degrade pollutants and clean up contaminated environments.

In summary, a deeper understanding of the inner electrical systems of the human body can inspire various applications in society by mimicking their unique properties and efficiencies. Bio-inspired technologies can revolutionize robotics, artificial intelligence, energy generation and storage, smart materials, communication, and environmental monitoring and remediation, among other fields. By leveraging the knowledge of the human body's electrical systems, we can create innovative solutions to address real-world challenges and improve the quality of life for all.

ABOUT THE AUTHOR

Ashai Parsons, an English scientist with a passion for exploring the dynamic nature of the universe, combines his interests in material sciences, mechanical physics, architecture, biochemistry, and quantum mechanics to create a series of thought-provoking books. By weaving together diverse theories of knowledge, he aims to stimulate intellectual exploration and foster creativity. His ultimate goal is to transform thoughts into art, paving the way for innovative ideas and envisioning new possibilities for the future.

www.ingramcontent.com/pod-product-compliance
Ingram Content Group UK Ltd.
Pitfield, Milton Keynes, MK11 3LW, UK
UKHW021933190726
13853UKWH00004B/1421

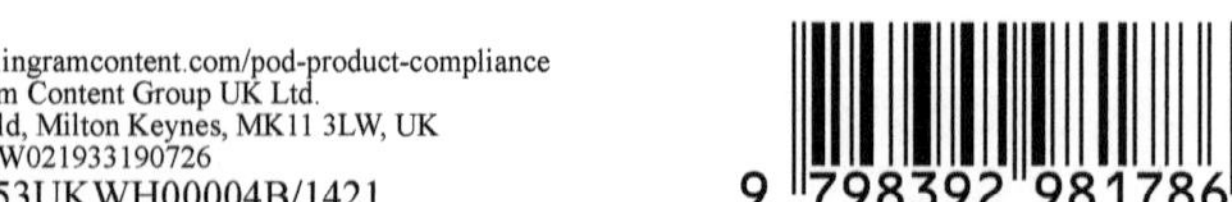